The Fantastic Book of Math Jokes: For Everyone, Not Just Mathematicians

UNCONVENTIONAL
PUBLISHING

www.unconventionalpublishing.com.au

Shane Van

ISBN : 978-0-6458679-6-1 PAPERBACK
ISBN: 978-0-6458679-7-8 HARDBACK
ISBN: 978-0-6458679-8-5 EBOOK

Preface

Unconventional Publishing proudly presents a book that will make you laugh and think about life differently. When I was asked to make a math joke book, I said, “I’ll make 1 if I halve 2.” Nothing in this book should be taken as fact (except the life of Pythagoras; that shit is fucked up) and is only here to put a smile on your face; no part of this book is meant to offend. If it does in any way offend you, you can always put it down and move on to something else.

This book is the next instalment in a series of *The Fantastic Joke Book*s and follows on from *The Fantastic Book of Hospitality Jokes*. This series roasts all sciences and everyday professions. The series will cover professions from chemists to physicists, doctors to plumbers. Hopefully, there will be some book that will suit your interests, if not just let us know, and we can create

Table of Contents

1
3
6
5
0
4
9
2
7
8
3
2

Basic Math

Why is the equal sign always so humble?

It realized it wasn't greater than or less than anyone else

Why was 10 sad?

Because 7, 8, 9

Why was 10 sad?

0.35836795

Cos (789)

Why did 7 eat 9?

You are meant to eat 3^2 meals a day

Why aren't jokes in base 8 funny?

Because 7 10 11

What did 0 say to 8?

Don't you think that belt is a bit tight?

Why did double 4 skip lunch?

They already 8

Why was 69 scared of 70?

They had a fight once and 71

How do you make 7 even?

Take the 's' away

How do you make the number 1 disappear?

Just add 'g'

Have you ever wondered what's odd?

Every other number

I heard a really scary math joke the other day but I am 2^2 to repeat it

A man walked into a bar and ordered 10 earthquake cocktails.

"Now that is an order of magnitude," said the barman

Did you hear about the number that lost a 1/3?

It was numb

How do you add up a mountain?

You summit

What do you call a number that can't stay still?

A roaming numeral

What was Hermione's sisters called?

Hermitwo and Hermithree

Not all math jokes are bad, just sum

Math jokes can usually be summed up according to their topics:

> Calculus jokes – they are derivative
>
> Trigonometry jokes – they are too graphic
>
> Algebraic jokes – they follow a basic formula
>
> Arithmetic jokes – they are very basic
>
> Statistical jokes – they have some outliers with the occasional pun

What does 2 + 2 equal?

7. You don't like my joke? Don't worry, I have 3 more

How does a mathematician say goodbye?

Cal-cu-later

Why did the chicken cross the Mobius Strip?

To get to the same side

There once was a mathematician who was afraid of negative numbers. He would stop at nothing to avoid them

What do you call a wizard who is good at math?
A mathemagician

What is a butterfly's favorite subject?
Mothematics

5 ants became friends with another 5 ants and decided to rent out a condo.
Now they are tenants
So numbers have never been my thirte

x^2 asked x^3 "Do you believe in god?"
After a moment of deep thought, x^3 said, "I am not sure, but I definitely believe in higher powers."

Why was the faction concerned about marrying a decimal?
They would have to convert

Did you know that 5 out of 4 people have problems understanding fraction jokes?

Why didn't Goldilocks drink the glass with 8 bits of ice in it?

It was 2 cubed

Why do teenagers travel in groups of 3s and 5s?

Because they can't even

Did you know who was responsible for bringing back Roman Numerals?

I for one

A farmer turned to his sheep dog and said, "Hey, dog, did you bring in the sheep?"

"Yep," said the sheep dog. "I got 30 of them."

"30?" asked the farmer. "I only have 27!"

"Oh, I rounded them up," said the sheep dog.

Apparently, 85% of the population don't know basic math. Thankfully, I'm from the other 25%

I just failed my Roman Numerals test. I couldn't remember how to write 1, 1000, 51, 6, and 500.

IM LIVID

For Christmas I got given mental arithmetic. It's the thought that counts

I made a homemade abacus. I used polo mints and threaded a string through. It really improved my menthol arithmetic

There are 3 kinds of people in the world: those who can count and those who can't

I hated school. I failed math so many times, I can't even count

The minus sign was just sitting there, really depressed one day, when the plus came along.
"Are you okay?" asked plus.
"I don't know. Are you sure I make a difference?" replied minus.
"I'm positive," said plus.

How many seconds are there in a year?
12. January 2nd, February 2nd, March 2nd….

Why did 1/5 go see a masseuse?
It was two tenths

Did you know the Irish are good at multiplication?
They are always Dublin

Did you hear about the time Katniss challenged 1 to a fight, who then brought 3, 5 and 7 as back up?
The odds were against her

There is a fine line between a numerator and a denominator

Algebra

There was once a snake handler who was trying to breed his 2 snakes. Nothing he was doing was working, so he called the local zoologist. She came out and viewed the cages and property. She picked up the snakes and inspected them. Turning them around, she simply said, "Hmmmm."

She then turned to the snake handler and said, "Okay, I saw a pile of chopped up logs by the side of the house. I want you to make a table out of them. Then I want you to put both snakes in one cage, then place the cage on the table."

"I don't get it. Why?" he asked.

"Just do it, and we will see."

So, he did. He made the table, placed the snakes on it, and sure enough, a few months later, he had a whole lot of little snakes slithering around the place. He was amazed, so he rang up the zoologist to say thank you and ask how.

"Well, you see," she replied, "you have a pair of adders, and everyone knows to multiply adders you need a log table."

Just remember, factorials was someone's attempt to make math more exciting

How do mathematicians propose to their girlfriends?
With a polynomial ring

Why were the ancient Romans so good at algebra?
They always knew X was 10

Why are pirates so good at algebra?
They never forget the C

Where do math fish live?
Indices

How do ghosts solve quadratic equations?
By completing the scare

What kind of fish do fishermen use to calculate the profit from their catch?
They use cod-ratic inequalities

Two surfers who hadn't seen each other in ages were meeting up on the beach.

"Hey, brah, I haven't seen you in ages," the first surfer says.

"Yeah, brah, I wanted to study marine biology, so I could learn more about the ocean," says the second surfer

"Really, brah?"

"Yeah, brah, but it was too much math. There was a lot of algae, brah."

How does a mathematician reprimand their children?

If I haven't told you n times, I've told you at least n+1 times

Why do mathematicians take long walks in the forest?

Because of all the natural logs

What type of bird can solve math problems?

Owlgebra

Vector was walking along Cartesian Avenue when it bumped into a sad and confused Scalar.

"What's wrong?" asked Vector

"I just have no direction."

What is BJ?

A bad joke

What is B + cJ?

A bad, complex joke

Why isn't a bad complex joke funny?

Because the humor is imaginary

Dear Algebra, you really need to solve your own problems

Dear Algebra, stop asking me to find your X. She's never coming back

Who came up with algebra?

An X-pert

What do you call 2 surfers who love math?

Algae-bros

Why was the Hyperbola surprised to find out they were sick?

They were asymptote-matic

Why don't atheists use exponents?

They don't believe in a higher power

(P+L)(O+T) = PO+PT+LO+LT And thus the plot was foiled!

I asked my German friend if he knew what the square root of 81 was.
He said Nein

What does algebra and Backstreet Boys have in common?
They both want you to tell them Y

$a^2 + b^2 = c^2$
$\frac{1}{2} \times c \times h$
a
b
h
c
$\frac{22}{7}$
V = a
c
b
a

Geometry

I tried to make a geometry joke once; it fell flat. I'm guessing it was too plane. Now I am back to square 1

It takes roughly 3.14 pastry chefs to make a Pi

Did you know that 3.14% of sailors are Pi-rates?

It's actually impossible to know exactly how much sugar to add to pumpkin filling.
It's because it's impossible to calculate Pi.

What is the coldest triangle?
The ice-soscles

Why did the math student fail geometry?
He thought it was pointless

What do you get if you divide the circumference of a pumpkin by its diameter?
A pumpkin Pi

Why did the obtuse triangle fail the math test?

It was never right

What is a mathematician's favorite tree?

A geometree

What do you get if you divide 22 sheep into 7 pens?

A shepherd's Pi

Why was 6 afraid of Pi?

I don't know; it was an irrational fear

I got a detention while learning about geometry; I really didn't enjoy the aftermath

What do you call a kettle boiling on top of Mount Everest?

Hypotenuse (high pot in use)

I hate all these Pi jokes; they just go on forever

Why didn't the map grids go to the school disco?

Because they were all squares

What kind of triangles are tortilla chips?

An i-salsa-les triangles

A teacher was giving students a word and then asked them to give a sentence about it. When she got to little Jimmy, she asked him to use the word 'geometry'.

Little Jimmy said, "Well, you see, miss, there was once this little acorn, and it rolled down a hill and got planted in some soil. There it grew and grew and grew. Then one day, it looked down at itself and said, "Gee, I'm a tree!"

There was once an artist who discovered a 2-sided shape. He submitted his findings to the International Geometry Society for verification. They didn't like the new shape and started arguing with him over it.

"Let bi-gons be bi-gons," he argued back.

What do mathematicians always take on camping trips?

A pair of axis

What does a mathematician order from McDonald's?

A plane cheeseburger

Why would you divide sin by tan?

Just cos

Where did square go after murdering triangle?

To prism

Why did Yoda fail geometry?

He couldn't work with the triangles; it was either do or do-not angles

Which knight came up with the idea of a round table?

Sir Cumference

What did the witch say when she lifted the curse?

Hex-a-gon

What shape is always waiting for you in the Department of Motor Vehicles?

A line

Why did the geometry teacher have so much trouble teaching the class?

He felt as if it was pointless, and they were going around in circles

How do you get from point A to point B?

You take an x-y plane

Is it a sin to make bad math jokes?

Cos if so, tan I'm sorry

What do you call a dead parrot?
Polygon

Why don't you argue with circles?
There is no point

In a distant land lies a triangular lake encircled by three kingdoms. The first kingdom is affluent and mighty, inhabited by wealthy and prosperous people and a mighty army. The second kingdom is more modest yet still possesses a fair amount of wealth and a sizable army. The third kingdom is impoverished and struggling, with barely any army to defend it.
Inevitably, the kingdoms eventually go to war over control of the lake, a valuable resource. The first kingdom dispatches 100 of their finest knights, all clad in the best armor and each accompanied by a personal squire. The second kingdom sends 50 knights, equipped with fine leather armor and a few dozen squires. The third kingdom sends its lone knight, an elderly warrior well past his prime, along with his personal squire.
The night before the decisive battle, the knights of the first kingdom drink and make merry, partying into the late hours. The knights of the second kingdom, though not as wealthy, still have their own supply of ale and drink late

into the night as well. In the third camp, the faithful squire takes a rope, slings it over the branch of a tall tree, and hangs a pot from it. He fills the pot with stew, and he and the old knight share a humble dinner together.

The next morning, the knights in the first two kingdoms are hungover and unable to fight, while the knight in the third kingdom is too old and weary to get up. In their place, the squires from all three kingdoms take to the battlefield. The battle rages on into the night, but when the dust finally settles, only one squire is left standing—the squire from the third poor kingdom.

It just goes to show that the squire of the high pot and noose is equal to the sum of the squires of the other two sides.

Is it a sin to make bad math jokes?

Cos if so, tan I'm sorry

What do you call a dead parrot?

Polygon

Why don't you argue with circles?

There is no point

What did the circle say to the tangent?
Are you touching me?

A rooster walks in a circle while smoking a joint. What is the name for the ratio of the circumference of the circle to its diameter?
Chicken pot pi

Why was triangle given the MVP in basketball?
It always made the 3 pointers

I enjoy algebra; I can do some calculus, but geometry is where I draw the line

In a pencil case, what piece of equipment is in charge?
The ruler

Why was the quadrilateral late for work?
He took the rhombus

There was once a math teacher who drew an angle on the board. It was 54 degrees. Then on the other side, she drew a 36-degree angle. Then suddenly, to the class's amazement, the angles started talking to each other.

"Good morning, 54, don't you have some very straight arms today."
"Why thank you, 36. But look at how strong your vertex is."

The students all looked at the teacher with amazement. She simply sighed and said, "I hate it when this happens. They just don't know how to spell. They are meant to be complementary."

Why can't your nose grow 12 inches long?
Because that would make it a foot

Mathematician: I have developed an irrational fear of the vertical axis
Psychologist: Why?
Mathematician: WHERE?????AAARRRRGGGHHH
Why are North Koreans so good at geometry?
Because they have the supreme ruler

There once was a geometry teacher who had 2 children. He named the first one A and the second one B. Their rooms were both on the same plane; however, the kids didn't like that and would constantly tell their parents about it. They were coplanars.

What is a geometry teacher's favorite drink?

Ovaltine

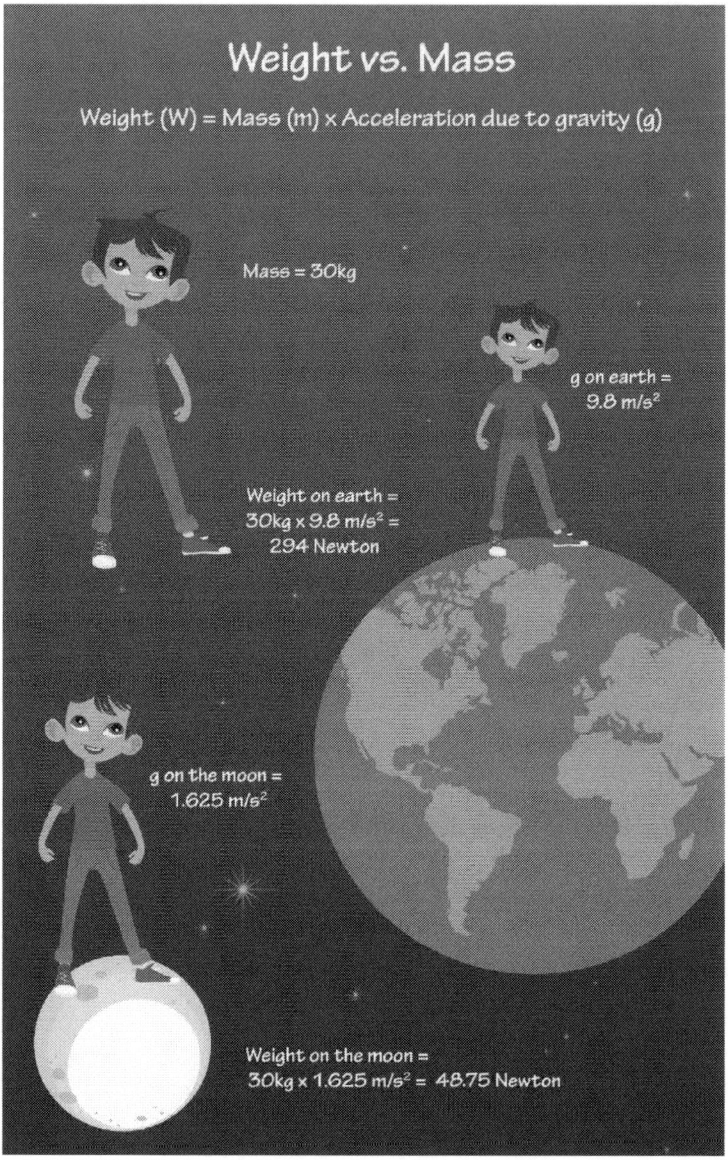
Weight vs. Mass
Weight (W) = Mass (m) x Acceleration due to gravity (g)
Mass = 30kg
g on earth = 9.8 m/s²
Weight on earth = 30kg x 9.8 m/s² = 294 Newton
g on the moon = 1.625 m/s²
Weight on the moon = 30kg x 1.625 m/s² = 48.75 Newton

Calculus

Why didn't the integral like going to the beach?

Because he didn't like the sun let alone the sun plus sea

What is the integral of (1/cabin) d (cabin)?

A logcabin

Why don't dentists like math?

They don't like calculus

There was once a guy with teeth so bad that his calculus had advanced to trigonometry

Who uses algebraic equations to work out the dimensions of coffins?

A math-e-mortician

I was never a huge fan of math; I had trouble differentiating it. It wasn't an integral part of my life and it just didn't add up

Why can't you differentiate social scientists?

Because they have no function

Why was the mathematician banging on a log?
He was studying logarithms

Why don't they teach calculus in the Middle East?
Because they don't like integration

To the people who don't like calculus, don't worry – your opinion will change over time

What do you call recycled calculus jokes?
Derivative humor

Calculus can't solve all your questions. It does have limits

Did you know when graphing data, the thickness of the data line is inversely proportional to the reliability of the data?

Teacher: Give me the integral of 1/cabin with respect to cabin
Student: A log cabin (laughing)
Teacher: No, it's a houseboat. You forgot the C

Why are frogs so good at calculus?
They always use der-ribbit-ives

Why can't the derivative of sec(x) go to the beach?

Because secant tan

I knew doing calculus would give me a heart attack… I should have seen the warning sines

Your momma is so fat, calculus still can't define the area under her curves

A mathematician was in his car doing 60 miles an hour when a colleague rang him.

"Hi, I'm 10 miles ahead of you and I'm doing 45 miles an hour."

"Look, I'll catch you later," he said.

Why is Ed Sheeran's favorite math unit on parabolas?

Because he's in love with the shape of U

What do you feed baby parabolas?

Quadratic formulas

Why are negative parabolas so introverted?

They have a hard time opening up

And on top of the Sermon Mount, Jesus said unto his disciples, “Truthfully, I speak to you with authority, I say unto thou, $y=x^2$”
The disciples listened then looked at each other in confusion and conferred until finally Peter asked, “Lord, we do not understand.”
“It’s a parabola, my son”

Soh Cah Toa

Trig

What is a math teacher's favorite salad?
Cos law

Why did the math teacher send the cold students to the corner of the room?
Because it was 90^0

I always struggle with understanding angles, but then one day, I did a complete 180

Why was sin lying on top of cos at the beach?
They were tanning.

What do you get if you cross a mountain lion and a mountain climber?
You mathematically can't; the climber is a scaler

A man set out to find the Holy Knights of Trigonometry, embarking on his journey after mastering all he could from his mentor. Eager to test his skills against one of the three legendary knights, he traversed vast plains, scaled treacherous mountains, sailed across dangerous seas, and even flew over a river of lava flowing from an ancient, long-dormant volcano. Suddenly, he noticed rain falling from a single point in the sky, curving as it descended toward the land. Following the rain, he ventured through a jungle and discovered that it flowed into a cave, where it cascaded down a hole at the cave's entrance. He threw a rock into the pit but only heard the sound of rushing water.

"Who goes there?" called a weary voice from the darkness.

"A man seeking a knight with extraordinary powers," he replied. "I've heard that three brothers were knighted by the king of this land and are known as the Trinity of Trigonometry. I hoped to find one of them here..."

An old man emerged from the shadows, clad in a toga that draped from his right arm to his left leg, leaving his left arm and right leg exposed. His once-dark beard had turned white and scraggly, obscuring most of his face, while his eyes conveyed the deep wisdom of one who had experienced much but was weary of life.

"So, you've come seeking a knight?" he rasped. "Well, you've found a man who was one, once. I am the middle brother. The eldest is Sindbad, and the youngest is Cosmos."

The man examined the old hermit with uncertainty and asked, "Are you Sir Tan...?"

I love my partner so much; she is the only sin (90^0) for me

Why couldn't the math student get a loan?

His parents wouldn't cosine

Math puns are the first sine of madness

Why did the obtuse angle go for a swim?

It was over 90^0 outside

Why are obtuse angles so sad?

They are never right

What is the best way to pass a math test?

To know all the angles

Why are circles overeducated?

They have 360 degrees

What do you call a heart-broken angle?

A wrecked-angle

I made a line graph showing all my past relationships. It has the ex-axis and the why-axis

An opinion without the 3.14159 is really an onion

10
A
10
A

Smart Guesswork

I sold my cabin out in the mountains to a Sasquatch.
He paid me in cryptid currency.

Why do sets of data only have 1 median?
If they had 2, they would be called co-medians

Did you hear about the latest stats joke?
Probably

I just got a new book on the theory of probability. What are the chances I will finish it?

Three statisticians went on a hunting trip when a pheasant jumped out at them. The first statistician took aim and fired. He missed by exactly 3 inches to the left. The second took a shot and missed by exactly 3 inches to the right. The third statistician started jumping up and down yelling, "We hit it, we hit it!"

Did you hear about the statistician who drowned trying to cross a river?
On average, it was only 2 ft deep

Optimist: The glass is half full
Pessimist: The glass is half empty
Statistician: The glass is double the necessary sample size

How do you make time fly?
Thow the clock out the window

Why isn't every fat man in a red suit with a white beard called Father Christmas?
Because correlation doesn't mean claus-ality

Correlation does not imply causality, but the more I say things like this at parties, the fewer amount of people want to talk to me

There are two types of people in the world;

1) Those that can extrapolate from incomplete data

For the three o'clock race, I backed a horse at ten to one. It came in at a quarter past four

I'm a bit worried about passing my probability exam. My chances are 30, 30

Math and Computers

Al gore actually started a punk rock band; he called it algorithm

What's the square root of Minecraft?
Beetroot, carrots, potatoes

What is the scariest type of algebra?
A boo-lean

Before computers, all the Boolean algebra was done by hand. Everyone hated it. It was boole sheet work.

Can we start calling calculators digital abacuses?

What is the difference between protractors and calculators?
When they start counting, no matter the angle, the protractors don't measure up

What do you get if you cross a duck with a calculator?
A quack-ulator

Why did Mike Tyson take a calculator to church?
He was invited to Thunday Math

I told my calculator a joke but it didn't compute the punchline

Did you hear about the broken fisherman's calculator?
There was something fishy which wasn't adding up

Did you hear about the new James Bond movie about a calculator?
It was called Casio Royale

Did you hear about the blonde who couldn't add 10 and 4 on the calculator?
She couldn't find the 10 key

How do you make a calculator happy?
Just keep calculating until you find the sum of happiness

Did you hear about the newer, faster calculator?
It's called a calcu-sooner

Did you hear about the newest high-speed calculator?
It's called a calcu-now

My minus button is missing from my calculator; it won't make a difference

Why was the calculator kicked out of the exam?
It kept trying to divide the class

Why was the calculator kicked out of the exam?
It wouldn't keep silent and kept adding to the conversation

Did you hear that Sony released a calculator that could also tell jokes?
It was a very pun-derful device

Why did the mathematician always bring a calculator to the party?
So he could always add some fun

What did the calculator say when it was complimented?
I am only doing my job

Why are calculators so popular?
They area

Use a calculator for the next joke
A woman went to the doctor for a checkup

Her waist measured 69 inches (*type 69)*

The doctor said, "Woah, that is too, too, too much!" (*type 222)*

He then gave her 51 diet pills (*type 51)*

In her panic, she took 8 times the amount (*multiple by 8)*

When she finished the dose and she looked down she became ….(*flip calculator)*

How do you get a calculator mad?

By pushing its buttons

Did you hear about the new talking calculators?

The results speak for themselves

My graphing calculator works great. Some would say it functions perfectly

Calculators may be ugly on the outside, but remember it's what on the inside that counts

When I was growing up, we were so poor I had a hand-me-down calculator. It was missing the multiplication button. Times were tough

Steve Jobs was once asked by a reporter why the iPhone was so expensive.

“It’s because it replaces a lot of devices,” said Jobs. “It’s a phone, a camera, a watch, music player, video recorder, voice recorder, GPS, map, calculator, a gaming console and a whole lot more. Why wouldn’t you expect to pay so much for one?”

“Why is the Android so much cheaper then?” asked the reporter.

“That’s simple. The Android only replaces 1 device – the iPhone.”

I once thought about getting a pocket calculator, then I realized that I actually don’t care how many pockets I have

I got a job working at a calculator factory. It isn’t my dream job, but all I have to do is add 1 button to the calculator. It’s a big plus

Orange
Water melon
Pineapple
FRESH FRUIT

Real-World Math

Mathematics don't get old; they just lose some functions

Two mathematicians and two physicists were going to a convention across the state. They decided to catch the train, and at the station, the physicists bought two tickets, but the mathematicians only bought one. The physicists just gave the mathematicians a funny look but didn't mention anything.

As the train ride progressed, the physicists saw that a conductor was going up the train checking for tickets.

"Hey, guys, the conductor is checking tickets," they warned the mathematicians.

The pair of mathematicians quickly got up and went into the bathroom to wait.

The conductor came up to the door, knocked, and yelled, "Ticket."

A train ticket slid under the door. The conductor took it and walked off. A few minutes later, the pair came out.

"That is brilliant, we will have to try that on the way home," said one of the physicists.

On the train ride home, the two physicists bought only one ticket, but this time, the mathematicians didn't buy a ticket at all. Sure enough, the conductor came back through looking at tickets. The physicists quickly got up and ran to the bathroom. A few seconds later, the mathematicians walked up to the bathroom door, knocked, and yelled, "Ticket."

Sure enough, the physicists slid the ticket under the door. The mathematicians grabbed it and ran into the next carriage toilet.

"Typical physicists. They use all our concepts without truly understanding the workings," said one of the mathematicians.

What does the Airforce and mathematicians have in common?

They both use Pi-lots

My pin is the last 4 digits of Pi

How is working at the fry station like studying Aristotle and Pythagoras?

You learn to appreciate ancient grease

How many ways are there to make a mathematician angry?

An infinite amount

Doctor: On a scale of 1 to 10, how bad is your pain?

Mathematician: Pi

Doctor: Sorry?

Mathematician: It's not too bad but it goes on forever

What does the moon and $1 have in common?

They both have quarters

If I had to describe myself in 3 words, it would be, "I am not good at math"

Why should you never talk about 288?

It's 2 gross

Did you hear about the kid who ate an abacus?

Turns out, it's what's on the inside that counts

A mathematician, a physicist, and an engineer were tasked with finding the volume of a rubber ball. The mathematician took the ball, measured the radius, then calculated the volume. The physicist submerged the ball in water; then, using Archimedes principles, calculated the volume. The engineer sat and studied the ball. He spun it round and round until he found the part number, and then he called the manufacturer.

I once hired this odd-job guy to help me around the office. Out of the 8 jobs he had to do, he only completed jobs 1, 3, 5, and 7.

How does a mathematician cut down a tree?
Ax-iomaticaly

Did you hear about the hen that counted her own chickens?
She was a mathemachicken

Monsters are not good at math – not unless you count Dracula

Why don't biologists and mathematicians get along?
Biologists believe multiplication and division are the same thing

Student: Do you do takeaways?
Chinese restaurant: We sure do
Student: What is 10178 – 359?

I lost my job in the abacus factory. They said I was counter-productive

I told my girlfriend that I had a crush on Beyonce.
She said, “Whatever floats you boat!”
“No, that’s buoyancy,” I said

My friend Simon drowned the other day. At his funeral, I made a wreath in the shape of a buoyancy ring. I know it’s what he would have wanted

There was a math convention just for blondes. The purpose was to break the typical stereotypes and provide more self-confidence.
During the first speech, a volunteer from the audience was brought on stage and asked simple math questions.
“What is 6 x 9?” asked the presenter.
“48,” said the volunteer.
The audience started shouting out, “GIVE HER ANOTHER CHANCE!!”
So, the presenter did.
“What is 12 + 14?” asked the presenter.
“23,” said the volunteer.
“GIVE HER ANOTHER CHANCE!!!” screamed out the audience.
“Okay, what is 2 + 2?”
“4,” said the volunteer.

"GIVE HER ANOTHER CHANCE!!!" screamed out the audience.

The reason why the Chinese are so good at math is because their dogs don't eat their homework

A blonde went for an interview with the police force. The officer asked a few questions just to ascertain her critical and spatial abilities.

Officer: What is 1+1?

Blonde: 2

Officer: What is 2 x 2?

Blonde: 4

Officer: What is the square root of 49?

Blonde: 7

Officer: Who killed Abraham Lincoln?

Blonde: I don't know.

Officer: That's a shame. How about you go home and think about it.

That evening at home, her friend asked her, "How did your interview go?"

"Great," she replied. "Not only did I answer everything correctly, but they already have me working on a case!"

Famous Mathematicians

Why was Albert Einstein and Pythagoras taking each other to court?

For the possession of c^2

Why did Pythagoras like math so much?

Just cos

Why was Pythagoras a murder suspect?

He knew all the angles

How did Pythagoras die?

He hung himself with a hypotenuse

How did Pythagoras die?

He died scalene a building

Who invented fractions?

Henry the $1/4^{th}$

Why are mathematicians always depressed?

They've had too many negative experiences

How can you tell if a mathematician is an extrovert?
They stare at your shoes in conversation

Pythagoras walks into a bar, muttering to himself, "On a right-angled triangle, the short side is X, the long side is Y, and the hypotenuse is Z. Then the square of Z must be equal to the sum of the square of X and the square of…of…um."
"Y the long face," says the bartender.

In an Indian camp beside a river sat 3 women on animal skins. One lady sat on a deer skin with her son. The son weighed about 60lbs. One lady sat with her son on a buffalo skin; that son weighed about 80lbs. Then one larger lady, who weighed about 100lbs, sat by herself on a hippopotamus skin.
The chieftain looked at this and realized the squaw on the hippopotamus was equal to the sons of the squaws on the other two hides.

Pythagoras is known for his work with triangles, but what isn't known is how he started a numbers cult and became irrational with Hippasus, who proved to Pythagoras that irrational numbers exist. Pythagoras was so irrational, he drowned Hippasus by holding his head under water. Later, a rival math cult burned down his home and killed his disciples. The rival cult chased him to a bean field, but because Pythagoras was terrified of beans, he didn't enter and got murdered outside. That or he got away and starved to death because he didn't like beans.

Why are mathematicians always depressed?
Because they are always filled with negative thoughts

How does a mathematician plow fields?
With a pro-tractor

Did you hear about the mathematician who invented 0?
He got thanked for nothing

Why did the mathematician get lost in the garden?
He was busy looking for square roots

Why did the mathematician always carry a ladder?
So he could figure out the higher-level problems

Why did the mathematician study astrology?
They wanted to find their sine

Why are mathematicians nicknamed math wizards?

They turn coffee into theorems

Why do mathematicians make good musicians?

They know how to count the beats

Why do mathematicians make good dancers?

They really know the algo-rhythm

What is it called when mathematicians retire?

The aftermath

What type of swimming stroke is a mathematician's favorite?

Dive-ision

Newton: I discovered calculus in 1664

Leibniz: No, I really discovered it 1670

Newton: Well, that is just derivative

How did Isaac Newton postulate calculus?

He went out on a Lim

Did you hear that some people thought Isaac Newton was gay?

Apparently, he wasn't integrals (into girls)

What does Mike Tyson call his graduate certificate in math?

His mathsters degree

A list of the most-watched mathematician movies:

Mean Girls

Sum Like It Hot

Deriving Miss Daisy

Gradients of the Galaxy

Pi Hard

Factor the Future

Along Came a Polygon

Sum Dog Millionaire

Quadrilateral Damage

Knock.

Knock.

Knock Knock.

Knock Knock Knock.

Knock Knock Knock Knock Knock.

Who's there?

Fibonacci

Did you hear about my Fibonacci sequence joke?
It's as good as my previous two jokes about the Fibonacci sequence put together

I have a friend who is really good at math. If we can't get the answer, we always see Tommy. Hilfiger it out

What was the bra size of Euclid's wife?
A. He only worked with flat surfaces

Rick was sitting in his bar in Casablanca, looking at and enjoying the extraordinary geometry of the wall patterns. He raised his glass to wall and said, "Here's looking at Euclid."

Archimedes had his buoyancy, Newton had his gravity, and Poincare had his hairy balls.

Why did Archimedes take a bath?
His wife told him, "You reek-a!"

What did Archimedes say when he pissed in the bath?
Urea!

I have a friend who was terrified of drowning in the sea. I tried to explain the Archimedes principle to him but he was too dense.

Archimedes' law of bathing: When the body is fully submerged in the water, a sibling will always need to pee.

There was once a horse that could do math. It became famous for it. Arithmetic was a breeze, and it learned it in no time. Algebra was simple. It even solved Euclidean geometry. When someone tried to teach it analytic geometry, it put up a fight, it reared its legs up, tried to kick and bite everyone, and became unruly.
It just goes to show, you can't put Descartes before the horse.

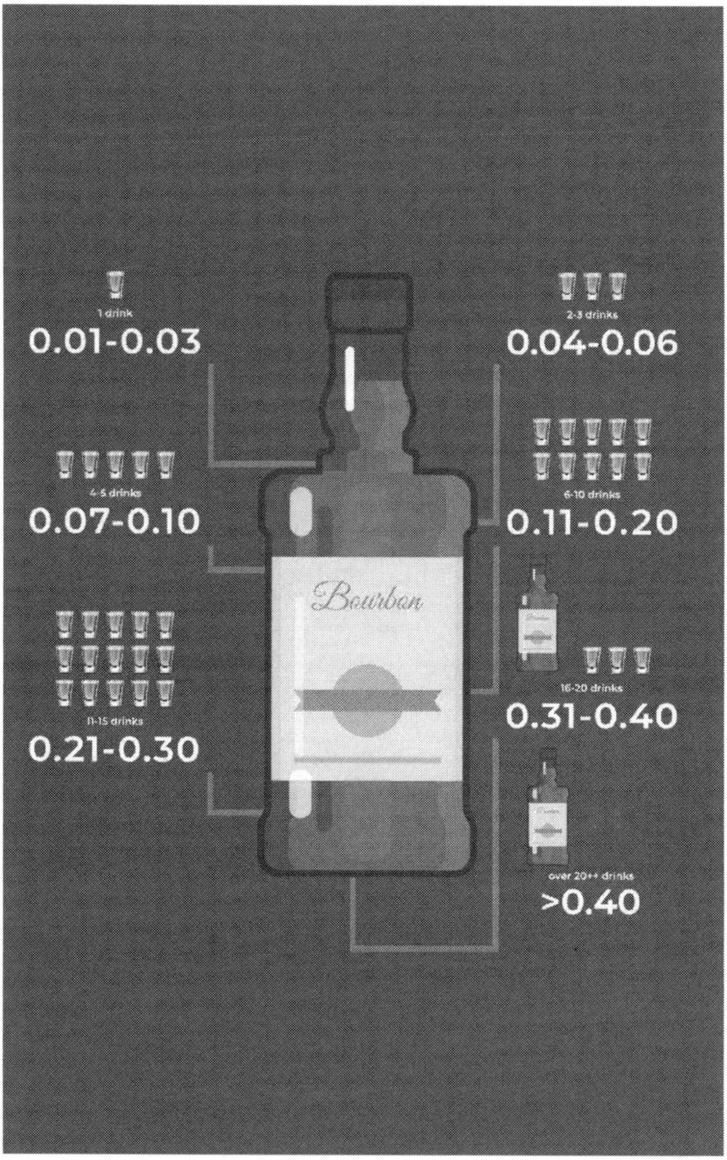
1 drink
0.01-0.03
2-3 drinks
0.04-0.06
4-5 drinks
0.07-0.10
6-10 drinks
0.11-0.20
11-15 drinks
0.21-0.30
16-20 drinks
0.31-0.40
over 20++ drinks
>0.40
Bourbon

Drunken Math

An engineer, a chef, and a mathematician went out for dinner and drinks. They ended up at their favorite steak house. While sitting there, a small fire erupted in the kitchen and the staff ran out in a panic.
The engineer did a quick calculation to see how much water he needed and then ran off looking for a bucket.
The chef saw that it was a grease fire, grabbed the cook, and told him to go get a fire blanket while he ran off and got a load of salt.
The mathematician looked at his friends, then to the fire, and then back to his friends. Realizing there was a solution, he just continued drinking.

Why wasn't 4 invited to the party?
Because he was 2 square

After drinking in a bar almost all night, a mathematician stumbled back home at 3 in the morning.
"Where have you been? You told me you would be back at 11:45," his wife yelled.
"Actually, I said I'd be home by a quarter *of* 12," he slurred.

Why did the mathematician get lost on the way home from the pub?
He couldn't figure out his 'add'-ress

What is a mathematician's golden rule?
Don't drink and derive

A mathematician walks into a bar and orders a root beer. "What's this, a square glass?" he complains. "Looks like I will just settle for a beer."

Where do mathematicians celebrate New Year's Eve?
Times Square
A drunk mathematician was walking past a shepherd who had lost control of his flock.
"Hey, I've got 68 sheep here and it's hard on my own. Can you help me round them up?" asked the shepherd.
"Sure," belched the mathematician. "70!"

Did you hear about the math professor who tried some bad amphetamines and thought he became a moth?
It was the old meth-math-moth myth.

Two mathematician students go to a university art night. At the function, they went to the bar, and the barman looked them up and down.
“We don’t often get your kind around here,” says the barman.
“What, mathematicians?” one of them says.
“No, math jokes.”

I went to a Chinese restaurant last night and ordered dishes 17 and 23. I sent them back as they tasted a bit odd
An infinite number of mathematicians entered a bar. The first one asked for a pint, the second one asked for a half pint, the third one asked for a quarter of a pint….
“Whoa, Whoa!” said the barman.” I know where this is going!”
He then filled 2 pints and slammed them on the table.
“You mathematicians just don’t know your limits,” he grumbled.

Adding in School

Why was Osama Bin Laden kicked out of math class?
He kept blowing up the pentagons

Did you hear about the student who murdered his math teacher with a calculator?
It was graphic

Saddam Hussein's old math teacher was arrested. He was carrying a calculator, protractor, and a ruler. He was charged with carrying weapons of math instruction.

How many teachers does it take to solve subtraction?
One…it only takes one teacher to make a difference

Did you hear about the math professor that was dreaming that he was giving a lecture?
He woke up and found out he was

Why was the calculus teacher bad at baseball?
He was good at fitting curves, not hitting them

Son: Dad, I need homework help

Father: No worries, ask away

Son: Dad, if I had 10 pineapples and I gave you 3, how many would I still have?

Father: How the fuck would I know? When I was in school, we worked out how to count watermelons!

Student: Teacher, did you know that 10 + 10 and 11 + 11 make the same?

Teacher: Actually, no they don't.

Student: Yeah, they do. 10 +10 equals twenty, and 11 + 11 equals twenty too.

Why don't you pick fights with a math teacher?

Because you are always outnumbered

There was a new foreign exchange student who had just arrived at school. This was the first lesson of the day, and it was math; in particular, it was fractions.

The student was having difficulty with the topic and so asked the teacher for help.

"Excuse me, Miss, I have trouble saying all these things in your language."

"Okay," says the teacher. "You say the top number first as the actual number or cardinal number, then you say the

bottom number as an ordinal number. For example, 2/3 you would say 2 thirds. Another one is 4/5 would be 4 fifths. How would you say 4/8?"
"Do you want me to make it smaller…um, I mean, simplify?"
"If you can," says the teacher.
"1 second."
"That's okay, take all the time you need," says the teacher.

Why did the calculus teacher confiscate the calculator?
The student was making graphic material

I saw the math teacher carrying graph paper. He must have been plotting something.

Why are math textbooks miserable?
They are full of problems

"I'm not normally late to math class," the math professor hastened to add.

"Algebra is stupid. Like seriously, when will I ever need to know?" a student complained to the teacher.
"You will never need it, but the smart kids in class will," replied the teacher.

What is a math teacher's favorite time of year?
Sum-mer

Where is the only place where you can buy 72 pumpkins and 12 monkeys?
In a math class

Why do all math professors wear glasses?
It improves division

Student: 67 544, 67 543, 67 542
Teacher: What do you think you are doing?
Student: I am determined to count backwards from 100 000 and I will stop at nothing.

Why did the math teacher open a bakery?
To teach the students about Pi's

Why did the math teacher open a bakery?
She wanted to show her students that division is a piece of cake

A teacher asks their students, "What is 2n + 2n?"
"I don't know," says little Jimmy. "It sounds 4n to me."

What is a math teacher's favorite snake?

A Pi-thon

Which snake is good at math?

An adder

What do math teachers eat?

Square meals

What do math teachers do when it snows?

They make snow angles

What is a math teacher's favorite tool?

Multi-pliers

Math teacher: Why the hell are you doing your work on the floor?

Student: You told us not to use the tables

Did you hear about the math teacher who confiscated a calendar?

Their days were numbered

Teacher: If you had 7 apples in 1 hand and 6 apples in another hand, what would you have?
Student: Huge hands

Why did the math teacher fire his gardeners?
Because his garden was filled with square roots

Three old university friends meet each other at a café. One was now a physicist, another a mathematician, and the last one a philosopher. They start jibbing at each other.
The mathematician turned to the physicist and said, "You know, physics is just applied mathematics!"
They had a bit of a laugh when the philosopher said, "And you know what? Mathematics is just applied philosophy!"
The physicist then said, "Shut the fuck up and finish making my coffee."

"Students are just getting dumber," complained the math professor. "The other day, I had a student ask if General Calculus was a Roman hero!"

If I got 50c for every math test I failed, I would have $6.30

The math teacher called me average. I thought that was very mean

A young boy came home from school and said to his dad, "Hey, Dad, my math teacher wants to see you."
"Why?" he asked.
"Well, I was in class, and the teacher asked me what is 4 x 6. I said 24. He then asked me, okay, now what is 6 X 4. I said well, what is the fucken difference?"
"Yeah, well, what is the difference?" said the father. "I will go and see him."
A few days later, the boy came home and went straight to his dad.
"Hey, Dad, did you speak to the math teacher yet?"
"No, I haven't!" said the dad.
"Well, Dad, now you gotta see the gym teacher as well."
"What for?" the dad said.
"Well, we are in the gym, and we are holding onto ropes. He tells me to lift my right leg, so I do, and then he tells me to lift my left. I then said, how do you want me to stand? On my cock?"
"I don't get it, how are you meant to stand? I will have to talk to him," said the boy's dad.
A few days later, the boy came home really upset, and the kid's father sees it.
"Hey, son, sorry I haven't had a chance to go and talk to your teachers yet."
"Don't bother, Dad. I got expelled today."
"What happened?" said the father mortified and angry.
"Well, I was asked to go the principal's office. When I got

there, the principal, math, gym, and art teacher were waiting for me."
"What the fuck was the art teacher there for?" interrupted the dad angrily.
"That's exactly what I said!"

Little Jimmy was sitting at the dinner table doing his math homework.
He was repeating to himself over and over again,"1 + 1, that son of a bitch is 2, 2 + 2, that son of a bitch is 4, 4 +4."
His mum walked past and heard him. "What the hell are you doing?"
"Oh, Mum, it's just our math homework. Our teacher told us to say it."
Furious, the mum rang up the teacher and gave her an absolute serving over the phone.
"It's absolutely disgraceful what is going on in your class. How dare you teach students to cuss and swear!!!"
"I didn't. What is little Jimmy saying?" asked the teacher.
"He is saying 1+1, that son of a bitch is 2!"
Laughing, the teacher said, "No, what I said was the sum of which."

A teacher was quizzing little Jimmy.
"If you had a quarter and then you asked your dad for another $1.50, how much money would you have?"
"None, Miss," said Jimmy
"You don't know basic arithmetic!" snapped the teacher.
"Nah, you don't know my dad," snapped Jimmy.

A biology teacher and a math teacher got together and decided to make a plan to save the planet. They called it the Al-gor-ithm.

A math student was really upset about their detention. It just didn't add up.

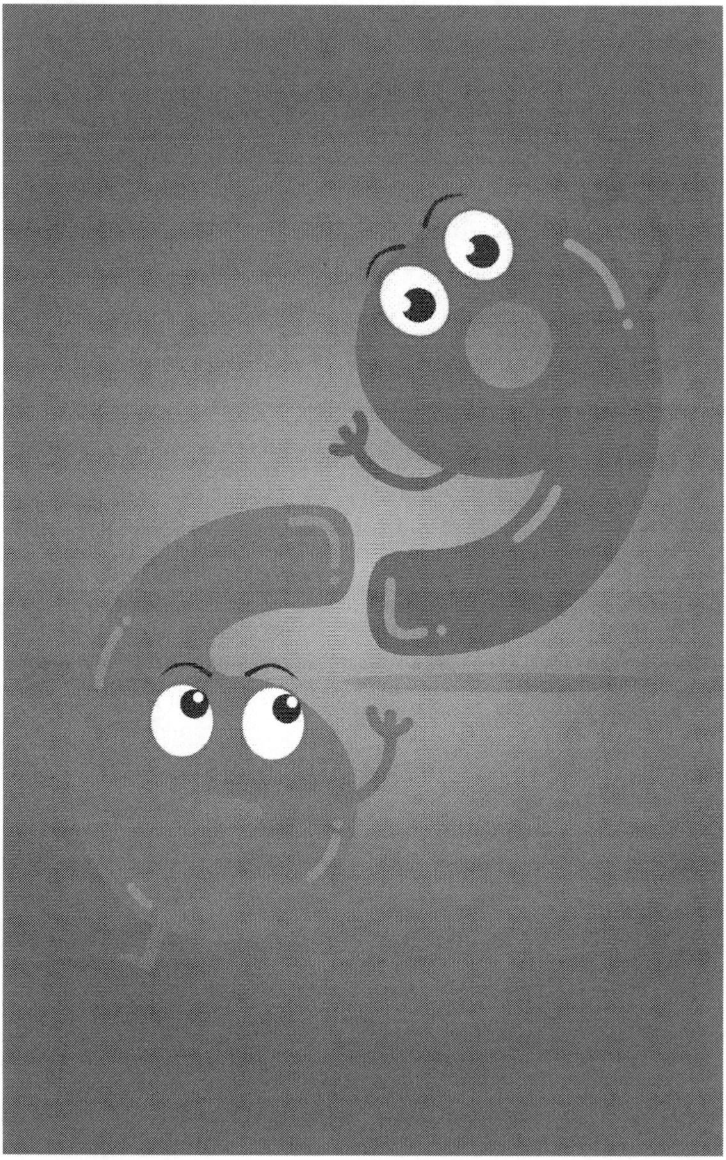

Numbers that Root

Two students end up failing their math exam and the next day, they end up going to their university professor and asking for a re-sit or anything they could do. The profession then suggested giving them 1 oral question right now, and if they can answer it correctly, he will allow them to pass.

So, the professor asks the first student, "You are sitting in train, and it gets too hot. What do you do?"

The first student replies, "I'll open the window."

"Good," says the professor "The window opening is 50 cm by 75 cm, the cabin's dimensions are 13m by 2.8m by 2.4 m, the train is traveling at 60km/hr. How long does it take to replace all the air in the cabin?"

The student just looked dumb founded and confused, then slumped their shoulders and said, "I don't know."

"Then you will still get the F," says the professor and dismisses him.

He then turns to the second student.

"Okay, now you are sitting in a train, and it gets too hot. What do you do?"

"I would take my jumper off," says the second student.

"It's still too hot. What do you do?"

"I would take my top off."

"It's still too hot, and you are starting to sweat a lot."

"I would just get naked."

"But, there is a guy staring at you now with an erection as you strip naked."

"Look, I'll let the whole train fuck me in the ass, but I will not open that FUCKEN WINDOW!" says the student.

Two historians were discussing ancient inventions.

Historian 1: Have you ever seen an Archimedes screw?

Historian 2: No. Wouldn't it be the same as all the other Greeks?

Two university friends were having a chat

"So, what happened to that cute girl you were seeing? The math major?"

"Oh, we broke up. She was cheating on me."

"What? How?"

"She was coming back late from the university bar, rang me up drunk, and told me she was in bed wrestling with 3 unknowns."

Trevor started taking night classes because he felt bad that he was never able to finish high school. Meanwhile, little Jimmy, his neighbor's kid, had started to truant school a lot, and his parents just couldn't get him interested. Trevor decided to try to get him interested in school again.

Trevor: "Hi, do you know who Victor Hugo is?"

Little Jimmy: "Nah."

Trevor: "He was the author of *Les Misérables*. I'm going to night school to learn this. If you went back to school, you would already know this."

A few days later, Trevor was leaving to go to night school again, and he saw little Jimmy out front.

Trevor: "Have you heard of Euclid?"

Little Jimmy: "Nah."

Trevor: "He was an Ancient Greek who was the father of geometry. I'm going to night school to learn this. If you went back to school, you would already know this."

A few days later, Trevor was leaving to go to night school again and he saw little Jimmy out front. Before he could say anything, Jimmy pipes up.

Little Jimmy: "Hey, have you heard of Andy Turner?"

Trevor: "Umm, no."

Little Jimmy: "He's the guy who's fucking your wife while you're at night school, maybe if you stopped night school, you would know this."

Little Jimmy goes home to tell his mum that he was suspended from school again.
“Well, what did you do this time?” his mum asks
“The teacher asked me, “If I gave you $25 and you gave $5 to Sarah, $5 to Lauren, and $5 to Alice, what would you have?..., Apparently 3 blowjobs and a kebab wasn’t an appropriate answer.”

What happens if you add a 1 into the calculator and then a 2?
You get a disgusting mess

Did you hear about the mathematician who became a porn star?
He changed his name to Cir-cum-ference

I have a fetish for algebra. I’ve come to terms with that

What do you call a horny geometry class?
An erectangle

Math is like sex; you don’t get extra marks for being fast

Parallel lines are the saddest math love story; so much in common but never be able to meet

What did the mathematician do when he was constipated?
He worked it out with a pencil

One day, the teacher during a math lesson asked, “Okay, Jimmy, 3 birds are sitting on a fence, and you shoot one. How many birds are left?”
“Oh, Miss, none ‘cause the other birds flew away!” said little Jimmy.
“No, Jimmy, there are 2 birds left because they flew away somewhere but I like the way you think,” the teacher replied.
“Okay, Miss, now let me ask you a question. 3 women are in a shop having ice cream. One is eating the ice cream, one is licking the ice cream, and one is sucking the ice cream. Which one is married?” little Jimmy blurted out.
The teacher rolls her eyes, “Okay, Jimmy, the one sucking the ice cream?”
“Nah, miss, it’s the one with a wedding ring on her finger, but I like the way you think.”

Why did the mathematician get divorced and marry his calculator?
He wanted someone he could rely on

Hamlet once had five children. He called the only boy Horatio. That’s a Horatio of one to 4.

What is the difference between women and algebra?
Hugh Hefner doesn't use algebra

My friend's nickname was geometry. When she became difficult, all the guys would cheat on her

Everyone is making jokes about radius, diameter, and Pi. It sounds like a big circle jerk

Men are like square roots: if they are under 25, they just do it in their heads.

Did you hear about the mathematician who was into polynomial relationships?
It didn't work out; there were too many variables

What is a mathematician's favorite boobs?
Quantitties

Sex is about the math: you add the bed, subtract the clothes, divide the legs, square root, and hope there isn't any multiplication

What do mermaids wear?
An algae-bra

Why was 6.9 sad?

It's a good time interrupted by a period

What is the square root of 69?

Idk 8 something

My girlfriend is the square root of -100. She is a perfect 10 but completely imaginary

In math class, when asked what comes after 69, "I do" is not an appropriate answer

There was a math party, and everyone was invited. The polynomials were making out with everyone, and square root was having a few drinks but was remaining in control. Exponential function was just standing to the side.
Inverse function asked, "What's wrong?"
Exponential function replied, "Even if I integrate, nothing will change."

Why did the mathematician get divorced?

He had too many improper relationships

Why are mathematicians so good in bed?

They don't divide their attention

A student was busy banging their calculator on the desk when the professor asked him, “What are you doing that for?”

“My calculator is broken, and I am trying to get it working,” said the student.

“Well, banging it on the desk will just break it. How would you like it if I banged you on the desk?”

My ex was called the human calculator because 14-year-old boys would do hand stands just to see her boobies.

Did you hear about the mathematician who swung both ways?

He was binomal

Why couldn’t the mathematician get a date?

He kept thinking of his *x’s* and y

A mathematician’s son asked, “Dad, what is an orgy?”

“Oh,” he replied, “it’s 230 divided by 3.3.”

A quite successful businessman is pondering over his bills when he looks up and asks his secretary, “If I were to give you $20,000 minus 14%, how much would you take off?”

“Everything except my earrings,” she replied.

Boy: Hey, beautiful, can I grab your number?

Girl: Sorry, I have a boyfriend.

Boy: Yeah, and I have a math test.

Girl: Okay, umm… why are you telling me this?

Boy: I thought we were mentioning things we cheat on.

Did you hear about the guy who blamed math for his divorce?

She finally put 2 and 2 together

What do you call it when you do math just for pleasure?

Mathturbation

There were two math teachers who were married for years. Things had gotten quite stale in their relationship, and they decided to bring in more excitement and try swinging. They ended up at a swinger's party and both had very different experiences.

The husband spent most of the night trying to flirt with a 21-year-old, which didn't go anywhere, while the wife met a fit 20-year-old and they disappeared into a bedroom for hours.

In hindsight, they should have known the outcome. After all, 20 goes into 48 more times than 49 into 21.

The sexy way how to remember Euclidean geometry:

- Transversal: the one that turns 2 singles into a threesome
- Vertical opposite angles: 1 glory hole connecting 2 of the same-size toilet stalls
- Alternate angles: Two angles with blue hair that hide in the same corner of their respective toilet stalls
- Co-interior angles: Glory holes on each side of the 1 toilet stall
- Corresponding angles: The glory hole is found in the same corner of 2 toilet stalls

Pick-Up Lines

Tinder is like geometry; if you have good lines, you will get good curves

Hey babe, what's your sine?

Hey babe, can you substitute my current girlfriend?

Hey babe, I love your well-defined function

Hey babe, why don't you put your hand down my pants and fill my exponential growth?

Hey babe, I'm like Pi – I'm really long and can go on without stopping

Hey babe, just think of me like your calculus homework – very hard and doing you on your desk

Hey babe, our love reminds me of Pi – it's irrational and never ending

Hey babe, do you know you make my heart beat faster than an airplane traveling 20m/s being pushed by a tail-wind traveling at 25 m/s?

Hey babe, I've memorized the first 30 digits of Pi, so your number should be easy

Hey babe, I am a mathematician so can I have your number?

Hey babe, can I just want to call-cu-later?

Hey babe, are you a square?
Because you have all the right angles

Hey babe, are you the square root of -100?
Because you're a solid 10 but you can't be real

Hey babe, are you an angle?
Because you're acute

Hey babe, are you a 45^0 angle?
Because you are perfect

Hey babe, are you the square root of -1?
Because you are too hot to be real

Hey babe, are you the square root of 2?
Because I feel irrational around you

Hey babe, I'm glad you aren't doing my differential equations homework
If you were, you would be 7 weeks late

Hey babe, are you good at math?
Maybe you can replace my eX without finding Y

Hey babes, you like math? (Yes or No, the answer doesn't matter)
Great, because the only number I care about is yours

Hey babe, do you like math?
How about adding my number to your contacts?

Hey babes, did you know I can get the square root of any number in 10 seconds?
What, don't you believe me? Give me your number and we can find out

Hey babe, are you a maths slut?
Because your legs are always divided

Hey babes, what does maths and my dick always have in common?
It is always hard and not everyone likes it

Wanna couple our equations tonight?

I love you so much it's like dividing by zero, it can't be defined

Can I have your significant digits?

Wanna expand my polynomial?

Damn, are you a calculus equation?
Because you are deriving me crazy
Roses are red,
Violets are blue
Math is hard
And so am I

Do you want to play math?

You be the numerator and I'll be the denominator and we both reduce to the simplest form

The derivative of my love for you is 0 because my love for you is constant

Damn, girl, are you math book? Because you're full of fucken problems

Damn, girl, I had the best math lesson. I was suppose find x, but I found u instead.

I just memorized the first 300 digits of Pi. If you give me the 7 digits of your number, I am sure I can memorize that too

You are sweeter than 3.14

Did you know the square root of all my fantasies is you?

I am so good at algebra that you and I would make 69

f(x)= x-1
y= x-1
x = y+1
= cx12
20T
axb=c
35°
A
1
B
3
2
C
H
G
E
F
D
A
B

Realistic Glossary

A

Abacus: An early calculator – in fact, your math teacher used one when he went to school

Absolute Value: The number of Absolut Vodkas you have before you pass out

Acute Angle: A very, very sexy angle

Addend: When you keep discovering bodies after a plane crash

Algebra: An algae's bra

Algorithm: The myth Al Gore keeps crapping on about

Angle: Not a being that has wings and comes from heaven

Angle Bisector: An angle that swings both ways

Area: This is my house

Array: A gun used by aliens

Attribute: The name given to 2 students, a boy and girl who have been selected for The Great Number games. The great 12 universities select 1 boy and 1 girl every year and send them to MIT. There, the students will fight to the death where the lucky last survivor will be allowed to live at MIT as a hero.

Average: Your ex

B

Base: What every good song needs

Base 10: When the clubs turns the base up to max

Bar Graph: A graph describing the local bars on price of drinks vs hot people present

Bedmas: What people fail to use when those stupid memes on Facebook appear. You quickly learn who dropped out of high school

Bell Curve: The shape of Bell's ass

Binomial: Someone who swings both ways and also has amazing abdominals

Box and Whisker Chart: The areas in the kitchen where you have moved the Whiskas box away from the cat, but it keeps getting them

C

Calculus: What happens to your teeth if you don't brush them enough

Capacity: How much you can drink in 1 sitting

Centimeter: The length of a penis

Circumference: It's not the length that women want

Chord: What art grads use to hang themselves when they discover they should have studied something useful like, I don't know, math!

Coefficient: A form of group work where one person does all the work and the other co-workers just coast along

Common Factors: If you look at the world and see the increase in violent crimes in areas, you will soon see some common factors

Complementary Angles: A very, very nice angle

Composite Number: A number that gets on with other numbers

Cone: What stoners use

Conic Section: Where the hose goes into the bottle

Constant: Your partner will never know what they want for dinner

Cylinder: Otherwise known as a circle-tube or snack can

D

Decagon: A priest of the holy order of Geometricus

Decimal: A tiny little dot that causes an identity crisis by ten fold

Denominator: The name of the Terminator movie in France

Degree: The arts one shouldn't really be one

Diagonal: Where Harry Polynomial purchased his magical abacus

Diameter: How a circle shows off how big it is

Difference: The awkward moment when two numbers don't agree and now they are too far apart to ever reconcile

Digit: An appendage of the human body that you can wiggle around. Is usually long. You can have fat ones or skinny ones. It's a finger, what were you thinking of, you dirty people?

Dividend: If you ever invest in shares, this should be the only thing you consider

Divisor: A French advisor

E

Edge: What guys do to try to make things last longer

Ellipse: What planets do

End Point: When edging fails

Equilateral: A communist lateral

Equation: Something that has an equals sign and has been confusing kids for hundreds of years

Even Number: A number that finally got its revenge

Event: A concert

Evaluation: The things that mortgage lenders go through to tell you that you – yes, you – can afford $500 a week in rent but not a $350-a-week mortgage.

Exponent: A defeated math opponent

Expression: The look on the face at the end point.

F

Face: Usually but not always consists of 2 eyes, a nose, lips, and a lot of teeth

Factor: The mysterious little helper who pops up to make sure your numbers behave properly

Factoring: The process of when a factor turns up uninvited to a party but then somehow divides the whole crowd into smaller, more manageable groups

Factorial Notation: When a factor takes notes in class

Factor Tree: A magical tree hidden in the forest where numbers go to find out who their factors were. Some get lost and lonely and turn into Primes

Fibonacci Sequence: How an Italian man figured out how Chinese whispers work

Figure: Someone's body shape, e.g., hourglass, athletic, chub stick or Roundy McRoller

Finite: My patience when asked the same question repeatedly

Flip: The middle finger still looks the same in the mirror

Formula: What chemists and chefs use. If a chef makes a mistake, it's not too bad; if a chemist makes a mistake, you get a fiery death

Fraction: A number that can't commit, always splitting itself into smaller parts

Frequency: How often do you hear your favorite song on the radio until after the 1000th time you are finally sick of it

Furlong: A mythical creature in World of Warcraft

G

Geometry: A topic that is deliberately confusing, 1 angle at a time

Graphing Calculator: A calculator that gives increasing violent and visual answers

Graph Theory: Mostly just a bunch of points ignoring each other until a line is connected with anyone

Greatest Common Factor: The strongest thing that brings two numbers together

H

Hexagon: A curse that a decagon removes

Histogram: Actually, it has nothing to do with history

Hyperbola: A bunch of lines trying to get away from each other

Hypotenuse: Someone smoking weed on top of a mountain

I

Identity: Numbers have no issue with this and don't need to make up a hundred more just to satisfy their own self-esteem

Improper Fraction: Fraction who likes to look up inappropriately

Inequality: What communists over exaggerate and try to use

Integers: Sometimes you don't want to know the number count

Irrational: The ex

Isosceles: Same same but different

K

Kilometer: A heavy meter

Knot: Kind of not

L

Like Terms: Gay terms

Like Fractions: Gay fractions

Line: Something you only cross when you want trouble

Line Segment: The safe place to cross

Linear Equation: An equation that is trying to be straight forward

Line of Symmetry: The VIP section of a shape where both sides agree to look exactly the same

Logic: Something that the world is lacking. Empathy is more important than logic, facts, or intelligence

Logarithm: The beat lumberjacks cut down trees to

M

Mean: An angry number

Median: A line on a map

Midpoint: Why the hell do you really need 3 different ways to describe the middle?

Mix Numbers: A bi-orgy

Mode: The current trend

Modular Arithmetic: The math version of a brock clock – it's always resetting itself and pretending nothing happened

Monomial: A lonely number

Multiple: Having a lot of partners

Multiplication: When the partners find out about each other (multi-complication)

N

Natural Numbers: Numbers that walk around without clothes

Negative Number: What you don't want from your bank balance

Net: A 2D shape that never has fish in it

N^{th} Root: How many roots do you get in a lifetime?

Norm: An insult used by woke people with blue hair

Normal Distribution: Looking at the amount of straight people during a Gay Pride Rally

Numerator: A number that has gone back in time to execute the denominator before division can occur

Number Line: The line that divides all the numbers into 2 groups: the sad, negative depressed downers on 1 side and the positive, go-lucky hippies on the other

Numeral: A fancy way of saying numbers, just like calling a sandwich a culinary masterpiece

O

Obtuse Angle: An angle that is up itself

Octagon: One of the easiest shapes to remember the number of sides

Odds: What you want in your favor when you're put in an arena where you hunt kids from other districts

Odd Number: A number that half the other numbers find weird

Operation: Something surgeons do

Ordinal: A rank in the mathly priest order

Order of Operations: How you play the game Operation

Outcome: The black eye you get when you try to pick up a lesbian girlfriend at a gay bar

P

Parallelogram: A word that is to try the linguistic skills of math teachers

Parabola: A paraplegic who caught Ebola

Pentagon: A building that was attacked during 9/11

Percent: Why do students not understand you can convert it to a decimal by dividing it by 100?

Perimeter: The leader of all meters

Perpendicular: When your dick gets caught in the vertical layer of mortar between two bricks

Pi: The name given to a number by a very hungry mathematician

Plane: What the terrorists used to crash into the Pentagon

Polynomial: A parrot that counts

Polygon: A runaway parrot

Prime Numbers: Optimus's siblings

Product: When a daddy mathematician and a mommy mathematician love each other very much, the children are called…

Proper Fraction: A fraction that has manners

Protractor: People who support the use of tractors in farming

Q

Quadrant: When four Karens complain

Quadratic Equation: When an equation has four mental problems

Qualitative: What happens when you give your quality friends laxatives

Quartic: How a geologist describes things looking like quartz

Quotient: What happens when the daddy mathematician and mommy get divorced

R

Radius: How fat your momma's ass is

Ratio: What you look for in the crazy-to-hotness scale

Ray: What aliens shoot you with. I am taking about actual aliens from other planets, not Mexicans or Middle Easterns

Range: What artillery bombardiers often get wrong

Rectangle: A wrecked triangle

Repeating Decimal: When the math is broken

Reflection: It depends on your mood. You can be handsome and have self-doubt and pick on your flaws, or you can be a 4ft hairy fat troll who gaslights everyone and thinks you're a prima donna or uomo

Remainder: When you accidentally leave one of your kids at the soccer field

Right Angle: The correct angle

Right Triangle: The correct triangle

Rhombus: What the Greeks call a roomy bus

S

Scalene Triangle: A confused triangle

Sector: A piece of pizza nobody wants (usually the vegetarian at the company's pizza party)

Slope: The nickname of Asians before it was politically incorrect

Square Root: When nerds have sex

Stem and Leaf: Hippies trying to do math

Subtraction: The removing of clothes

Supplementary Angles: Having lots of….a surplus of angles.

Symmetry: Where mathematicians get buried

T

Tangent: A side story that gets you sidetracked

Term: What teachers count down until the Christmas holidays

Tessellation: Star constellation that was named after Nicola Tesla

Translation: What things get lost in

Transversal: A versal that switches its gender

Trapezoid: It's the rebel of shapes; it doesn't follow the parallel laws like that nerd square. It's like rectangle but skips leg day

Tree Diagram: What hippies like to draw

Triangle: An angle that keeps trying no matter what

Trinomial: The nominal amount of people in mathematician's fantasy

U

Unit: Ronnie Coleman

Uniform: Ronnie Coleman when he still worked for the police

V

Variable: What a man calculates when trying to find a car space but his partner totally ignores him and questions his parking choice

Venn Diagram: The pretty circular way to say, "Same same, but different"

Volume: What you need to turn down, so you can see properly

Vertex: What little kids try to create in a swimming pool by everyone running/swimming around in a circle in the same direction

W

Weight: What you never ask a woman

Whole Number: A chubby number

X

x-axis: The lefty/righty line

x-intercept: An upgraded Star Wars spaceship

X: Where the treasure is buried

x : The random unknown number that changes whenever you discover it

Y

Y-axis: The upy/downy line

Yard: The length you always miss the hole by in golf

Z

Zero slope: No Asians on that hill

Author's Note

Hi there.

Thank you for taking the time to read this book. Coming from a science background, I have been wanting to write these joke books for a long time and unleash the inner nerd in me. I hope you had just as many laughs as I did in writing this book. If you enjoyed reading this, please click on the link to leave a review or where you bought this book from. That would be greatly appreciated.

You will also find another joke book on the website, one that is not for sale, one that is not for the fainted heart, one that can be considered a bit risqué and politically incorrect. Consider this a gift for taking the time to purchase and read my book, but be warned – only download it if you are not easily offended.

www.unconventionalpublishing.com.au

It was mentioned in the introduction that this book will be part of an upcoming series. If science jokes tickle your fancy and you haven't had a chance to read the first book, look at *The Fantastic Book of Chemistry Jokes.* There will also be a clothing line with some of the one-liners on them. If you want to see any other professions being roasted or want a particular joke on a shirt, please visit our website and let us know.

Kind Regards

Shane Van

Other Books in the series

Printed in Dunstable, United Kingdom

213cff1f-7943-4986-a976-2fe6d871f46dR01